What others are saying about this book:

"I can honestly say that I have never seen such an effective method for teaching multiplication and division. I myself struggled with a learning disability while growing up and I found this method very easy to comprehend. If this had been available to me, my struggles would have been quickly resolved."
~ Rachel R. Godfrey, parent

"Sticks and Steps® is a concrete, sequential method for teaching students of all ages how to multiply and divide. It is quick and **fun**! Math concepts can seem very abstract, but in just a short time, not only will Sticks and Steps® help students achieve a better understanding of mathematics, it will boost confidence and self esteem as well." ~ Cheryl Woodwell, Elementary Educator, Gifted Certified

"What a wonderful resource for children who experience problems with multiplying and dividing. It is the easiest and most understandable method I've ever used." ~ Donna Boss, grandmother to 14

"Multiplication was so much easier after learning Sticks and Steps.® I felt so important when my teacher asked me to teach it to others." ~ Chase Lyttle, student

"Here is a valuable mathematical resource for any parent or teacher trying to instill an 'I can do this!' attitude in their child or student. Ms. Stearns is a teacher who has a knack for identifying methods that help students easily succeed. Her Sticks and Steps® method is easy to master and offers a win-win situation for all."
~ Nancy Ellington, M.A. Ed., National Board Certified Teacher

"As a 15 year teacher working with struggling students, I am always in search of a method that will 'click' for children having difficulties with multiplication and division. The Sticks and Steps® strategy has enabled my students to conquer skills they once thought were beyond their reach! Sticks and Steps® is a valuable resource for all educators to carry with them in their bag of tricks."
~ Theresa Leinbach, Early Intervention Specialist for Exceptional Children

Multiply and Divide with Sticks and Steps®

Teach this Easy Method in Just 5 Minutes

Maureen Stearns
Exceptional Student Educator, B.A., M.S.

ENRICHMENT
B·O·O·K·S

Websites:
www.enrichmentbooks.com
www.sticksandsteps.com

Publisher's email: publisher@enrichmentbooks.com
Author's email: maureen@sticksandsteps.com

Sticks and Steps® is a trademark of Maureen Stearns.

ISBN-10: 0-9726908-5-9
ISBN-13: 978-0-9726908-5-0
Library of Congress Control Number: 2010921215

Design by Janice Phelps Williams: www.janicephelps.com

PRINTED IN THE UNITED STATES OF AMERICA

Stearns, Maureen.

Multiply and divide with sticks and steps : teach this easy method in just 5 minutes / Maureen Stearns. -- 1st ed. -- St. Petersburg, Fla. : Enrichment Books, c2011.

p. ; cm.

ISBN: 13-digit: 978-0-9726908-5-0 ; 10-digit: 0-9726908-5-9
Summary: Intended for anyone having difficulties learning how to multiply or divide, this clearly illustrated text shows how to teach multiplication and division using a special technique.

1. Multiplication--Study and teaching (Elementary) 2. Division--Study and teaching (Elementary) 3. Arithmetic--Study and teaching (Elementary) I. Title.

QA115 .S74 2011   2010921215
372.7/2--dc22   1006

In Loving Memory

Denise Lynn Otteni

April 26, 1973 – March 3, 2010

A book holds a house of gold.
~ Chinese Proverb

To Parents:

It's you against your kid…

It's nine o'clock at night. You expect the evening struggle to begin — homework is not done. Your child says, "We're learning about multiplication in class and *I … don't … understand … it!*"

Usually about now you both begin to scream or cry in frustration, but this time it's different. The new book you've picked up seems nothing short of spectacular.

If you have knowledge, let others light
their candles at it.
~ Margaret Fuller (1810–1850)

To Teachers:

The stakes are high. You know your students' mastery of multiplication can make or break your math assessment scores. You've over-drilled their multiplication tables. *"My students just can't seem to learn!"* you hear yourself saying. The next day a note says, *"Billy cried last night — he just doesn't understand multiplication."* Now it's up to you to find a better way … students and parents … are counting on you …

Congratulations!

You have in your hands an innovative and simple way to teach multiplication and division. Sticks and Steps® is a multi-sensory method that is easy to learn, easy to teach, and easy to remember.

Even though there are many ways you can use Sticks and Steps, the basic method is the core of this book. Best of all, it takes only 5 minutes to teach. Because I wanted to keep information short and simple with respect to your time, I did not make this book thick with work pages that could seem like information overload.

The concepts here can be easily used with any multiplication or division workbook used today.

Enjoy the book !

Maureen

The Beginning of Sticks and Steps®

My job as an exceptional student educator is to be on the lookout for better ways to help struggling students. Several years ago, I worked with a fifth grade student who could not memorize her multiplication facts, let alone understand the concept of multiplication. She would resort to drawing tallies on paper hoping to get the answer. It was an embarrassment to her, and it was not fail-proof. Often, she counted tallies incorrectly and ended up with too few or too many. Despite her best effort, she was frustrated — and so was I.

My search for a better way began. The goal was to decrease my student's tally counting errors. I also wanted my student to arrive at the answer quickly and not be hampered by her inability to memorize. With the method I developed, my student became very proficient with arriving at the answer quickly and easily. She ended the year with a smile on her face, knowing she mastered a special technique that her peers did not know. I called this method Sticks and Steps and I soon started to teach Sticks and Steps to my other students. I was amazed by how fast they caught on, and so the method was launched. Even my younger students now learn multiplication. If they can count, they can multiply — and get the RIGHT answer every time!

Why It Works

Multiplication can be thought of as layer upon layer of the same amount, added together. In order to master multiplication, it's beneficial for struggling students to find a multiplication strategy that's logical, tactile, and visually understandable. This multi-sensory method does not rely on memorization but rather on the ability to touch and count.

As students gain confidence with multiplication concepts, they will naturally want to jump ahead by skipping step counts they already know. Eventually, they may not need to use this method at all. But whenever answers to multiplication facts slip their minds, they know they can pull Sticks and Steps 'out of their back pockets' to use at any time.

A good teacher is like a candle — it consumes
itself to light the way for others.
~ Author Unknown

Table of Contents

Easy to understand ... Simple to remember

Learning is a treasure that will
follow its owner everywhere.
~ Chinese Proverb

Chapter One

The Basic Method

(No time for tears here!)

Let's begin.

These vertical lines will be referred to as "sticks."

These horizontal lines will be referred to as "steps."
(one below the other)

The number of sticks and the number of steps will vary depending on your multiplication factors.

A.

Let's find the answer to:

3 x 4

Draw 3 sticks, then 4 steps.

3 sticks

The first number is a 3. This means you will draw 3 sticks.

The second number is a 4. This means you will draw 4 steps right next to the 3 sticks.

4 steps

B.

Solving 3 x 4

Using the tip of your pencil, touch and count each stick (start at the left).

Begin here:

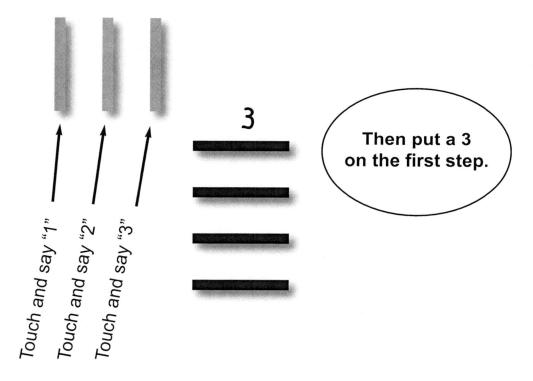

When you teach your son,
you teach your son's son.
~ The Talmud

C.

Solving 3 x 4

Say "3," then touch each stick and count again.

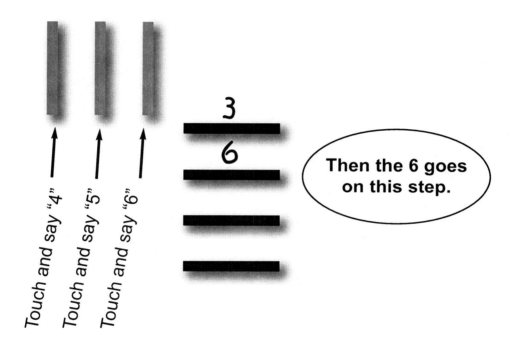

3
6

Touch and say "4"
Touch and say "5"
Touch and say "6"

Then the 6 goes on this step.

D.

Solving 3 x 4

Say "6," then again touch and count through the sticks.

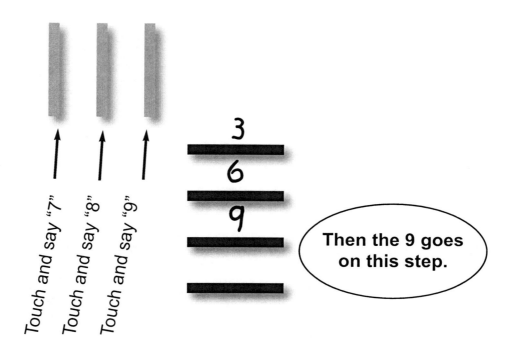

Touch and say "7"

Touch and say "8"

Touch and say "9"

3

6

9

Then the 9 goes on this step.

E.

Solving 3 x 4

Say "9," then touch and count the sticks once more.

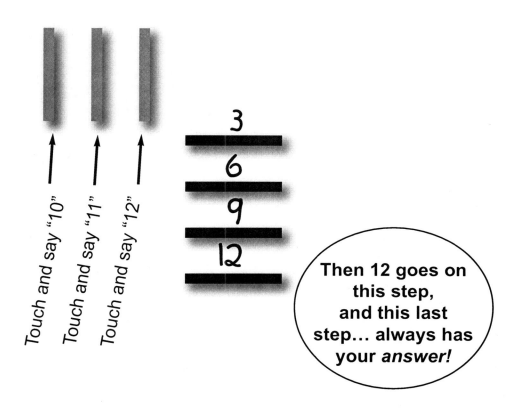

3 x 4 = 12

If you can count, you can multiply and
get the right answer every time!

Nothing is particularly hard if
you divide it into small steps.
~ Henry Ford (1863–1947)

Sticks and Steps Sampler

3 x 3 = 9

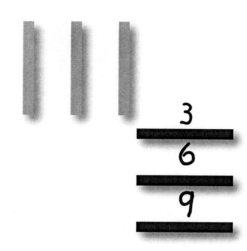

4 x 6 = 24

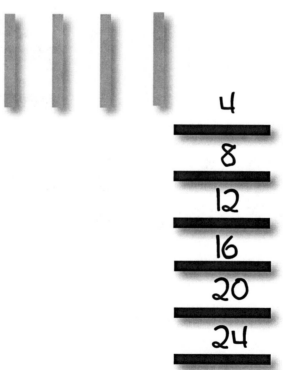

3 x 9 = 27

3
6
9
12
15
18
21
24
27

9 x 5 = 45

9
18
27
36
45

4 x 8 = 32

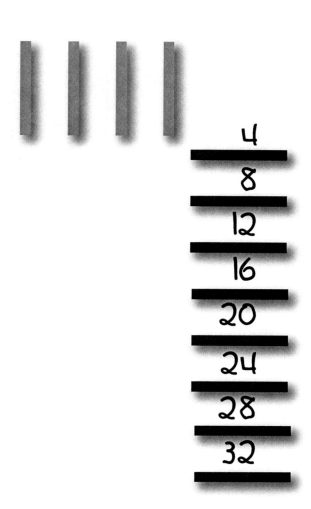

7 x 2 = 14

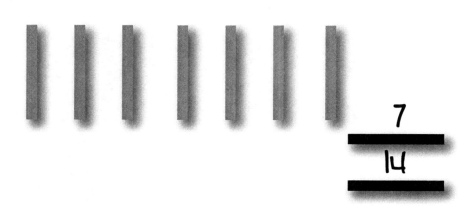

8 x 7 = 56

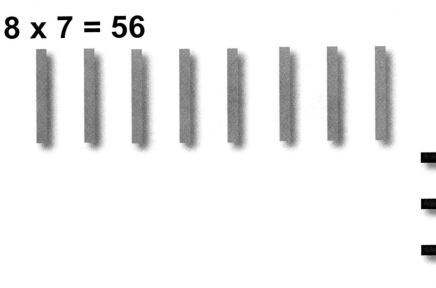

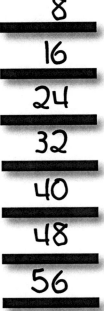

8
16
24
32
40
48
56

6 x 3 = 18

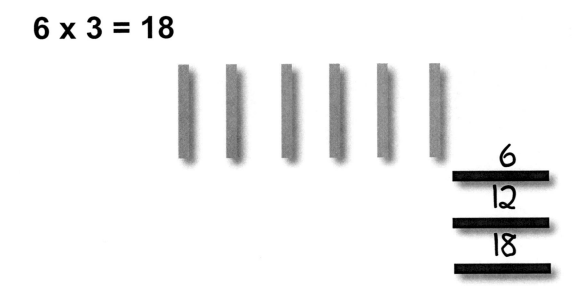

6
12
18

2 x 9 = 18

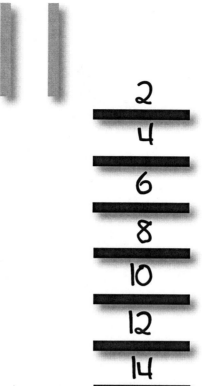

12 x 3 = 36

To waken interest and kindle enthusiasm is the
sure way to teach easily and successfully.
~ Tryon Edwards (1809–1894)

Chapter Two

Swap the Sticks with the Steps

A.

Let's swap these numbers (also referred to as "factors").

Will the answer be the same? Let's try it.

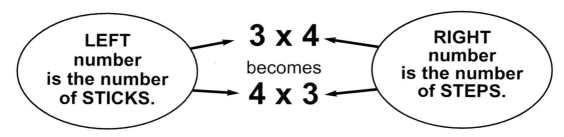

This time draw <u>4</u> sticks

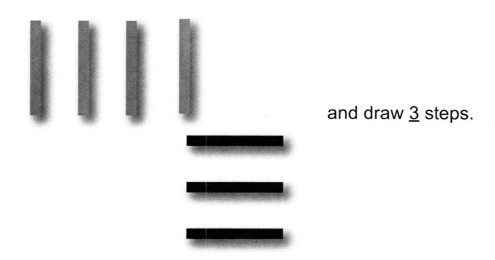

and draw <u>3</u> steps.

B.

Solving 4 x 3

Touch and count each stick.
You have 4.

<u>Write 4 on the first step.</u>

Say "4" — then touch and count each stick once again.
You now have 8.

<u>Write 8 on the second step.</u>

Say "8" — then touch and count each stick again.
You now have 12.

<u>Write 12 on your last step.</u>

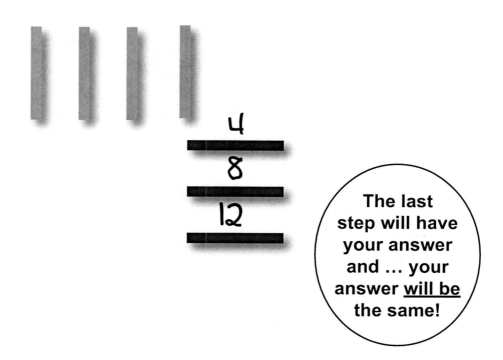

The last step will have your answer and … your answer <u>will be</u> the same!

It does not matter if you swap factors (your sticks with your steps, or your steps with your sticks) because your product – another name for the multiplication answer – will be the same.

C.

Here's another example …

4 x 6

6 x 4

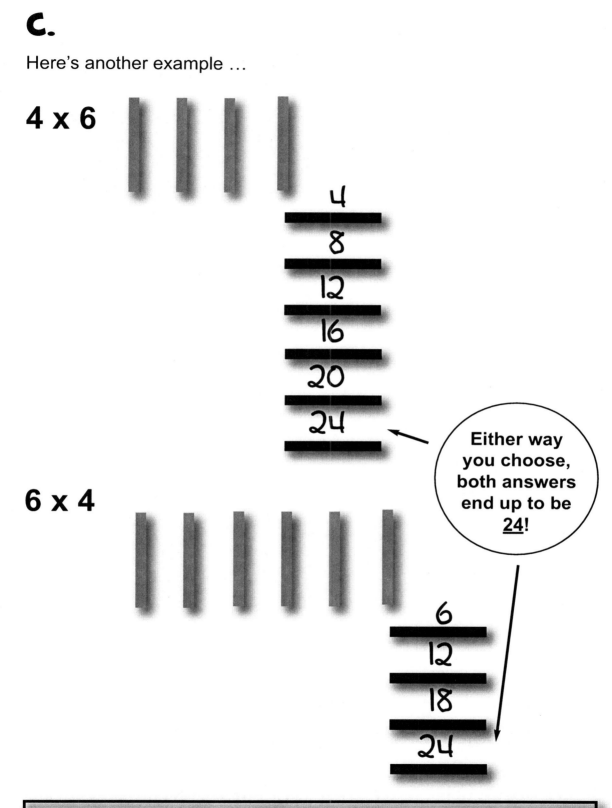

This is called the **commutative property** of multiplication.

The object of teaching a child is to enable him
to get along without his teacher.
~ Elbert Hubbard (1856–1915)

Chapter Three

"Oh no! ... A factor is missing."

A.

What to do when … you <u>know</u> how many **sticks**…

$$3 \times ? = 24$$

but <u>don't know</u> how many **steps,**
but you know your answer will end up to be 24.

First do this Draw <u>3</u> sticks.

(You won't know how many <u>steps</u> to draw right now.)

Next do this Count through the three sticks. Stack your number counts, one below the other, as shown.

3

6

9

12

15

18

21

(STOP! You have reached 24.) 24

B.
Finally do this...

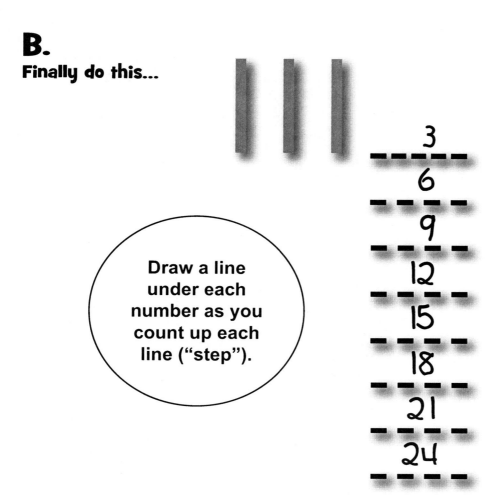

Draw a line under each number as you count up each line ("step").

3
6
9
12
15
18
21
24

How many steps to get to 24?

It took 8 **steps**, which is the unknown factor, 8.

$$3 \times \boxed{8} = 24$$

Learning is a kind of natural food for the mind.
~ Marcus Cicero (106 BC–43 BC)

C.

What if the first factor is unknown?

$$\_\_ \times 3 = 24$$

Just swap your factors and you are on your way.

(Use the same instructions in A and B on the previous pages.)

$$3 \times \_\_ = 24$$

Remember … swapping factors is permitted because multiplication is <u>commutative</u>.

You will always need to know your <u>stick</u> count in order to do Sticks and Steps multiplication (or division).

So <u>swap</u> your <u>factors</u> whenever you need to.

Chapter Four

"Oh look! I see a pattern."

With Sticks and Steps®, you can find other answers within answers.

Example: Below, the last step shows the answer for 6 x 9, which is 54 (six sticks, nine steps). But within these steps are answers for *many* multiplication facts.

All the numbers in the steps column are referred to as "multiples."

Here you see multiples of **6** — a layering upon layering of sixes.

6 = 6 x 1 (<u>six</u> sticks, <u>one</u> step)

12 = 6 x 2 (six sticks, <u>two</u> steps)

18 = 6 x 3 (six sticks, <u>three</u> steps)

24 = 6 x 4 (six sticks, <u>four</u> steps)

30 = 6 x 5 (six sticks, <u>five</u> steps)

36 = 6 x 6 (six sticks, <u>six</u> steps)

42 = 6 x 7 (six sticks, <u>seven</u> steps)

48 = 6 x 8 (six sticks, <u>eight</u> steps)

54 = 6 x 9 (six sticks, <u>nine</u> steps)

Now that you know this ... see the next chapter.

Knowledge is power.
~ Sir Frances Bacon (1561–1626)

Chapter Five

Multiply Multi-digit Numbers

It's easy when you use Sticks and Steps.

Solving

$$162$$
$$\underline{x\ \ 4}$$

First, find the biggest digit in the top row. It's a "6." So then, make four sticks and six steps.

Since 6 is the largest digit in the top row, **six** steps will be the most steps you will need to draw in order to solve this problem.

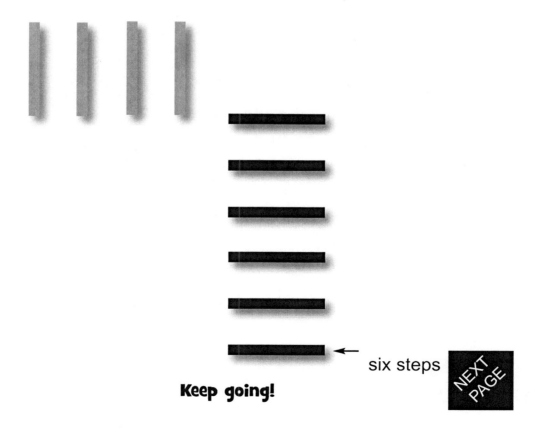

six steps

Keep going!

NEXT PAGE

51

Solving

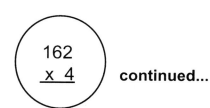

162
x 4

continued...

A.

> **Note: In order to multiply multi-digit numbers, knowledge of <u>place value</u> and <u>regrouping</u> (sometimes known as "carrying") will be needed.**

Now, fill in all of your steps counting groups of four:

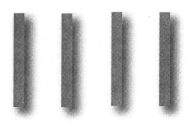

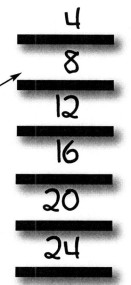

To find the answer for 4 x 2 (your first pair of numbers to multiply), go to the <u>second</u> step.
Find the answer: 8.

The 8 goes below the line, in the ones column.

162
x 4
8

Next, multiply 4 x 6. Go to the <u>sixth</u> step to find your answer: 24.

Write 4 below the line in the tens column, regrouping the 2 by writing it above the 1 that is in the top row, the hundreds column.

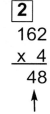

2
162
x 4
48

Solving

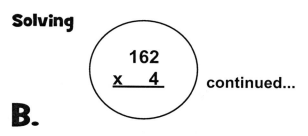

162
x   4

continued...

B.

Next, multiply 4 x 1.

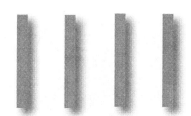

Go to the first step, find your answer: 4.

| 2 |
| 162 |

Add 2 (that was regrouped) to 4, which then equals 6.

x   4

648

Put the <u>6</u> below the line in the hundreds column.

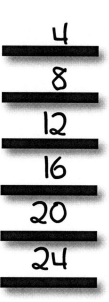

4

8

12

16

20

24

Conclusion:

Any numbers needed for solving 162 x 4 were found on one of the steps above.

Education is the mother of leadership.
~ Wendell L. Wilkie (1892–1944)

C.

Even with solving a multiplication problem this large, you will only need **seven** steps, 7 being the largest digit in this top row.

$$2,134,2\boxed{7}3 \text{ (steps)}$$
$$\times \ 4 \text{ (sticks)}$$
$$\overline{8,537,092}$$

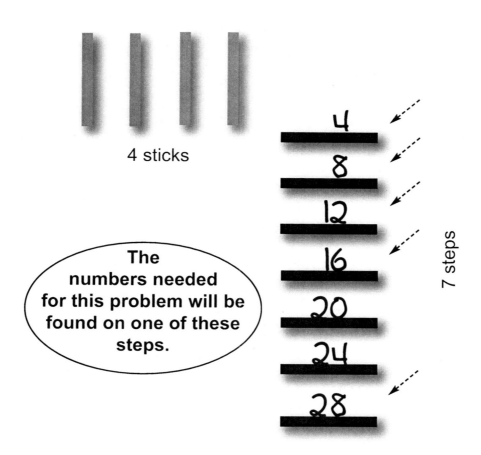

4 sticks

The numbers needed for this problem will be found on one of these steps.

4
8
12
16
20
24
28

7 steps

D.

How many steps will you need to draw for these examples?
(Remember, it will be the **largest** digit on the top row.)

6,74⟦8⟧,125 (steps)
    x 4 (sticks)

How many steps will you need to draw for this problem? (Hint . . . it's 8.)

1,436,45⟦9⟧ (steps)
    x 5 (sticks)

How many for this one? (It's 9.)

⟦5⟧,423,423 (steps)
    x 4 (sticks)

And for this one? (It's 5.)

Important: When multiplying <u>multi-digit</u> numbers, DO NOT switch your sticks with your steps or you will "mess up." Remember, the <u>number you are multiplying by</u> will always be the sticks.

Chapter Six

Take a Shortcut

As you learn how to multiply with the Sticks and Steps® method, you will discover shortcuts. Here's an example:

Solving 4 x 8

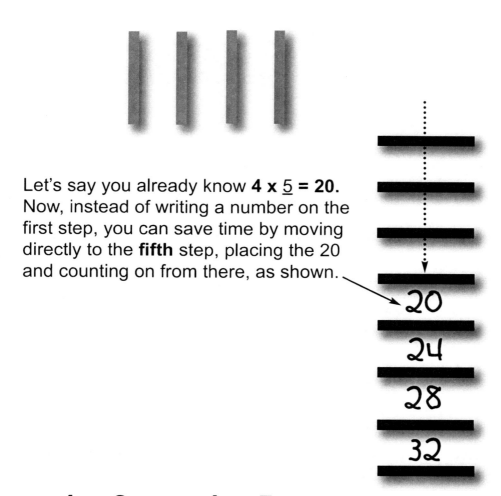

Let's say you already know **4 x 5 = 20.** Now, instead of writing a number on the first step, you can save time by moving directly to the **fifth** step, placing the 20 and counting on from there, as shown.

20
24
28
32

So, **4 x 8** is really **4 x 5** (which equals **20**), PLUS **3** more groups of **4**.

What we have to learn to do, we learn by doing.
~ Aristotle (384 BC–322 BC)

Chapter Seven

Divide Using Sticks and Steps®

Division is the inverse (or opposite) of multiplication.

For example: 5 x 4 = 20 is the inverse of 20 ÷ 4 = 5

$$5 \times 4 = 20 \qquad 20 \div 4 = 5$$

When the same three numbers appear in both multiplication and division, they are called "fact families."

In **Chapter Three** you learned what to do when the step count is unknown, as with $3 \times ? = 24.$

To review:

- ☞ count through the three sticks,
- ☞ stack your number counts, one below the other,
- ☞ stop at 24,
- ☞ then **underline and count** the <u>number</u> of steps: your <u>answer</u>.

Division works the same way.

Your division answer (also called "quotient") is just the unknown **step** count when you are multiplying.

$$5 \times ? \text{ (steps)} = 20 \qquad \textbf{same as} \qquad 5 \overline{)20}^{\;?}$$

A.

Let's Practice Dividing

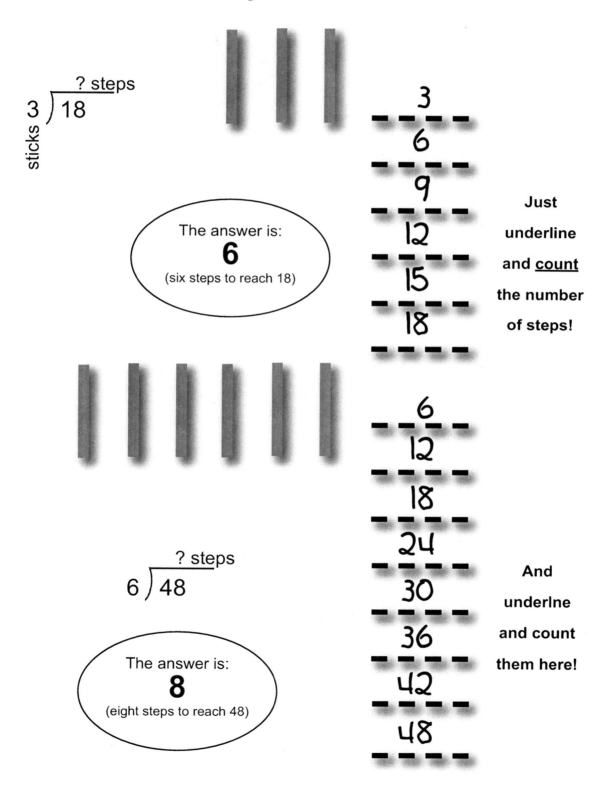

sticks 3) 18 ? steps

3
6
9
12
15
18

The answer is:
6
(six steps to reach 18)

Just

underline

and <u>count</u>

the number

of steps!

6
12
18
24
30
36
42
48

6) 48 ? steps

The answer is:
8
(eight steps to reach 48)

And

underline

and count

them here!

I hear and I forget. I see and I remember.
I do and I understand.
~ Chinese Proverb

Chapter Eight

Divide Large Numbers

When students learn long division, it's no wonder the process can produce tears (as it did for me long ago!). Students must learn how to switch their thinking from dividing to multiplying to subtracting, to bringing down a digit and then repeating this process all over again. Sticks and Steps® helps to simplify and organize this thinking process.

??? steps

$$4 \overline{) 2748}$$

☞ First do this:

Start with basic division (4 sticks) and count through the sticks until you reach 27 or less. Why? Because 2 and 7 are the first two digits in the division box (mathematically referred to as the "dividend").

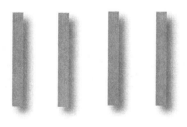

Now count the steps.
You have 6.

See the next page...

4
8
12
16
20
24

67

Put the 6 right above the 7 that's in the division box.

$$4 \overline{)\underline{27}48} \quad ^{6}$$

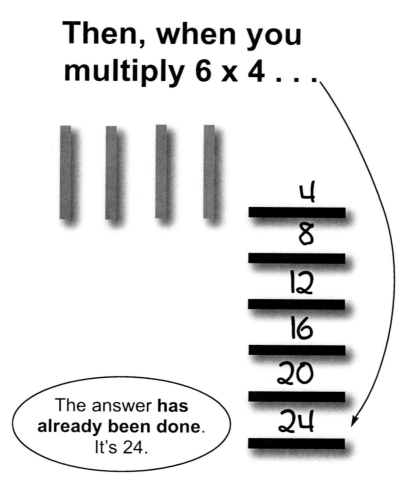

Then, when you multiply 6 x 4 . . .

4
8
12
16
20
24

The answer **has already been done**. It's 24.

$$4 \overline{)2748} \quad ^{6}$$
$$\underline{-24\downarrow}$$
$$\boxed{34}$$

Now, subtract 24 from 27.

Then bring down the digit 4 to sit beside the 3.

Education is the kindling of a flame,
not the filling of a vessel.
~ Socrates (470 BC–399 BC)

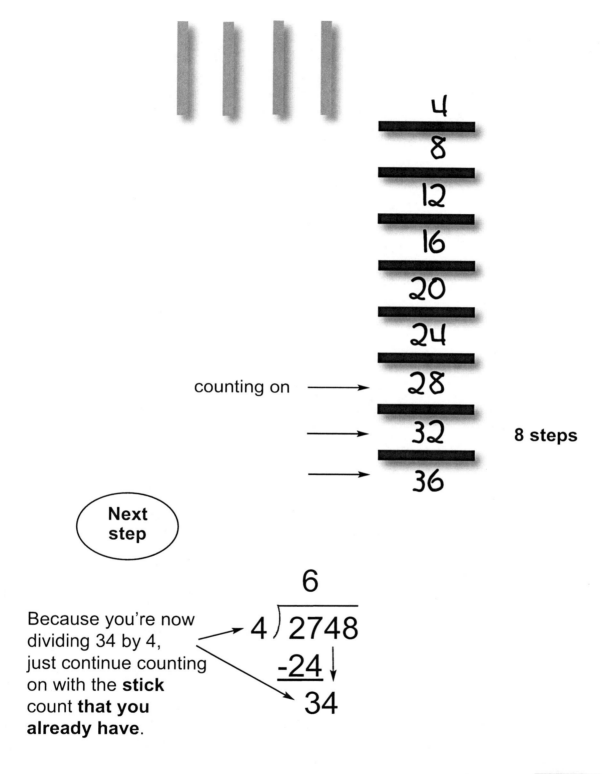

counting on ⟶ 28

⟶ 32   **8 steps**

⟶ 36

4
8
12
16
20
24
28
32
36

**Next
step**

6

4 ⟌ 2748

-24

34

Because you're now
dividing 34 by 4,
just continue counting
on with the **stick**
count **that you
already have**.

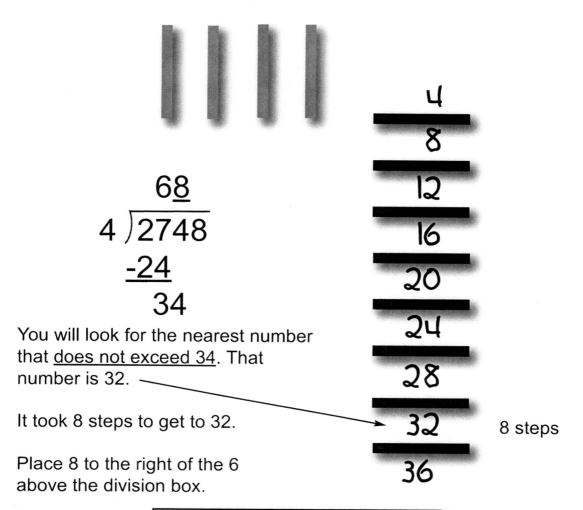

$$\frac{68}{4\,)\,2748}$$
$$\frac{-24}{34}$$

4
8
12
16
20
24
28
32   8 steps
36

You will look for the nearest number that <u>does not exceed 34</u>. That number is 32.

It took 8 steps to get to 32.

Place 8 to the right of the 6 above the division box.

The process just repeats

Multiply 8 x 4.
Put 32 below 34.

$$\frac{68}{4\,)\,2748}$$
$$\frac{-24}{34}$$
$$-32$$

➡

Subtract 32 from 34

$$\frac{68}{4\,)\,2748}$$
$$\frac{-24}{34}$$
$$\frac{-32}{2}$$

NEXT PAGE

Bring down 8.
Place it next to the 2.

Divide 28 by 4.
Refer to the stick count
You already drew
(on the previous page).
It took 7 steps of 4.
Place 7 next to the 8.

$$\begin{array}{r} 68 \\ 4\,)\overline{2748} \\ -24 \\ \hline 34 \\ -32 \\ \hline 28 \end{array}$$

→

$$\begin{array}{r} 687 \\ 4\,)\overline{2748} \\ -24 \\ \hline 34 \\ -32 \\ \hline 28 \end{array}$$

Next step

Multiply 7 x 4.
The answer is found on step 7 (previous page).
No amount remains.

With long division, it's difficult to
remember the order of operations:
   divide
      multiply
         subtract
            bring down
               repeat.

$$\begin{array}{r} 687 \\ 4\,)\overline{2748} \\ -24 \\ \hline 34 \\ -32 \\ \hline 28 \\ \underline{28} \\ 0 \end{array}$$

Sticks and Steps® makes long division easier!

It is the mind that makes the body.
~ Sojourner Truth (1797–1883)

Chapter Nine

Find the Least Common Multiple (LCM)

First, you will need to know what the term "multiples" means and then the term "common multiples."

Multiples are obtained just by multiplying a specific number by other whole numbers, such as 1, 2, 3, 4, 5, 6... Looped below are some multiples for the number "4."

$$4 \times 1 = \boxed{4}$$
$$4 \times 2 = 8$$
$$4 \times 3 = 12$$
$$4 \times 4 = 16$$
$$4 \times 5 = 20$$
$$4 \times 6 = 24$$

Because 4 is a factor that makes up numbers such as 4, 8, 12, 16, 20, 24, these numbers then become "multiples of 4."

You can also find the multiples of 4 by adding 4 to 4 again, and again.

$$4 + 4 = 8$$
$$4 + 4 + 4 = 12$$
$$4 + 4 + 4 + 4 = 16$$
$$4 + 4 + 4 + 4 + 4 = 20$$
$$4 + 4 + 4 + 4 + 4 + 4 = 24$$

A.

The Term "Common Multiples":

Here is an example of finding multiples for two numbers:

multiples of 3:	multiples of 4:
3 x 1 = 3	4 x 1 = 4
3 x 2 = 6	4 x 2 = 8
3 x 3 = 9	4 x 3 = **12**
3 x 4 = **12**	4 x 4 = 16
3 x 5 = 15	4 x 5 = 20
3 x 6 = 18	4 x 6 = **24**
3 x 7 = 21	4 x 7 = 28
3 x 8 = **24**	4 x 8 = 32
and so on	*and so on*

Common multiples refers to the same multiples that show up in each column. As you can see, the numbers 12 and 24 appear in both columns, which make them common multiples.

Now, the **Least Common Multiple** is the smallest common multiple that appears in each column. Here it is 12. The number 24 is a common multiple, but not the *Least* Common Multiple.

When would you need to know the Least Common Multiple (LCM)? When you are adding or subtracting fractions with different denominators.

Note: The **denominator** is the bottom number in a fraction.
In this example, $\frac{1}{3} + \frac{2}{4}$, the denominators are 3 and 4.
The **Least** Common Multiple will be called the *Least* Common Denominator (LCD) when you are specifically working with fractions.

He who opens the school door,
closes a prison.
~ Victor Hugo (1802–1885)

B.

With Sticks and Steps®, finding your **Least Common Multiple** is easy. For multiples of 4 and 5, for example, simply do this:

For multiples of 4,
make four sticks.

For multiples of 5,
make five sticks.

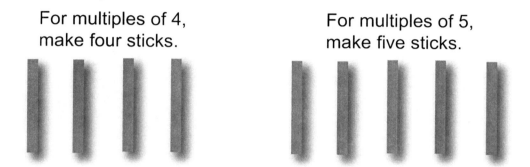

Knowing your number of steps is not important here. Just start counting your sticks over and over, then enter each number, one below the other.

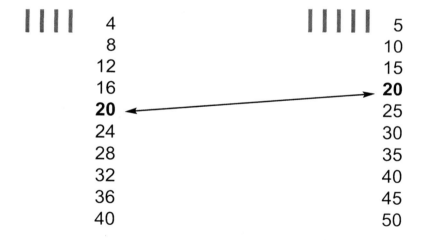

4	5
8	10
12	15
16	**20**
20	25
24	30
28	35
32	40
36	45
40	50

The LCM or Least Common Multiple is 20. It's the smallest number that shows up in **both** columns. Can you see it?

For finding the LCM for more than two numbers, see the sampler on the next page.

C.

Least Common Multiple Sampler

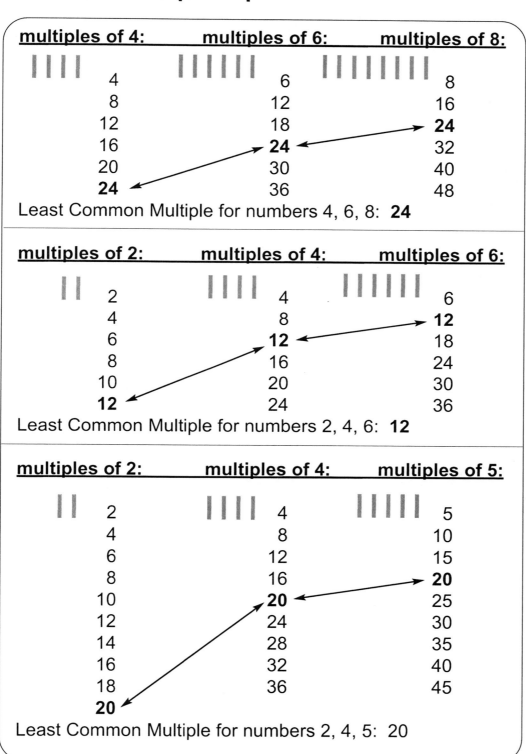

multiples of 4:	multiples of 6:	multiples of 8:
4	6	8
8	12	16
12	18	**24**
16	**24**	32
20	30	40
24	36	48

Least Common Multiple for numbers 4, 6, 8: **24**

multiples of 2:	multiples of 4:	multiples of 6:
2	4	6
4	8	**12**
6	**12**	18
8	16	24
10	20	30
12	24	36

Least Common Multiple for numbers 2, 4, 6: **12**

multiples of 2:	multiples of 4:	multiples of 5:
2	4	5
4	8	10
6	12	15
8	16	**20**
10	**20**	25
12	24	30
14	28	35
16	32	40
18	36	45
20		

Least Common Multiple for numbers 2, 4, 5: 20

By learning you will teach;
by teaching you will learn.
~ Latin Proverb

Chapter Ten

Easy Doubles and Square Roots

A "double" is what your answer is called when you multiply two identical numbers together such as:

$$2 \times 2 = 4$$
$$3 \times 3 = 9$$
$$5 \times 5 = 25$$
$$7 \times 7 = 49$$

doubles

With Sticks and Steps®, finding numbers that are considered doubles is easy. Anytime your number of sticks <u>and</u> your number of steps is the same, your answer (the number on last step) will be called a "double."

Doubles Sampler:

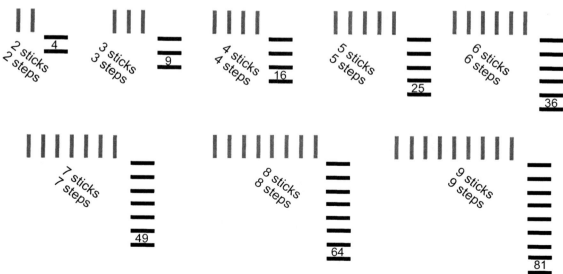

Go on to next page.

A.

So, what's the big deal about doubles?

Understanding **doubles** will help you understand the concept of "**square roots.**" A square root is the special number that, when multiplied by itself, equals a given number referred to as the "double." For instance, the special number 3, when multiplied by itself, gives you the double, 9: (3 x 3 = 9). This special number 3 is called the "square root of 9."

The mathematical symbol for square root is this... $\sqrt{}$

...and "the square root of 9 is 3" looks like this: $\sqrt{\dfrac{3}{9}}$

With Sticks and Steps®, whenever your **stick** count and your **step** count are the **same**, that same number is the square root of the number appearing on your last step. For example:

2 is the square root of 4.

$\sqrt{\dfrac{2}{4}}$

Two sticks and two steps,
with 4 appearing on the last step.

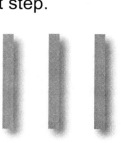

3 is the square root of 9.

$\sqrt{\dfrac{3}{9}}$

A teacher affects eternity; he can never tell where his influence stops.
~ Henry Adams (1838–1918)

5 is the square root of 25.

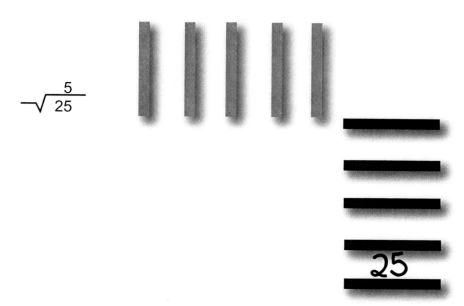

$\sqrt{\dfrac{5}{25}}$

7 is the square root of 49.

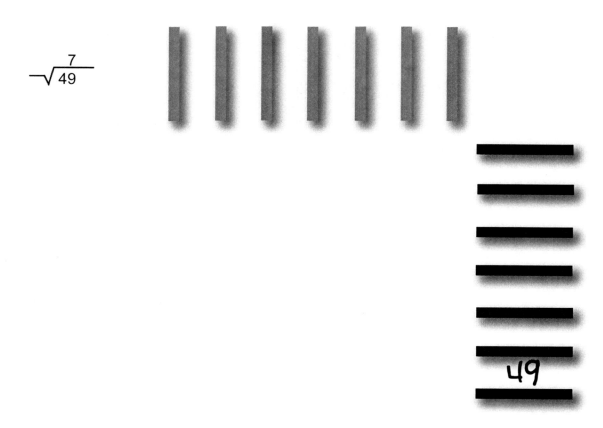

$\sqrt{\dfrac{7}{49}}$

If you think about it, using Sticks and Steps® makes perfect sense because you can create a "<u>perfect square</u>" around the sticks and steps like so:

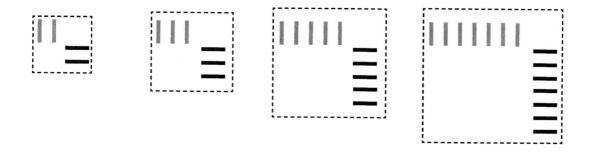

Thanks to my nephew,
Grant, for pointing out
to me this "perfect square."

Chapter Eleven

Special Rules to Remember

Anytime you multiply using a zero, the answer will always be zero. This is called *"the multiplicative property of zero."*

Some examples:

$$2 \times 0 = 0$$
$$8 \times 0 = 0$$
$$7 \times 0 = 0$$
$$4 \times 0 = 0$$

In fact, even the multiplication problem below equals zero...

$$3 \times 4 \times 5 \times 6 \times 2 \times \underline{\mathbf{0}} = \underline{\mathbf{0}}$$

...because <u>a</u> <u>lot</u> of <u>nothing</u> is *<u>nothing</u>*!

Also, whenever you multiply two numbers together and one of the numbers is a "1," the <u>other</u> number will always be your answer. This is called *"the identity property of multiplication."*

Some examples:

$2 \times 1 = 2$		$1 \times 2 = 2$
$8 \times 1 = 8$		$1 \times 8 = 8$
$7 \times 1 = 7$		$1 \times 7 = 7$
$4 \times 1 = 4$	or	$1 \times 4 = 4$
$5 \times 1 = 5$		$1 \times 5 = 5$
$6 \times 1 = 6$		$1 \times 6 = 6$

Give a man a fish and you feed him for a day.
Teach a man to fish and you feed him for a lifetime.
~ Chinese Proverb

The Ultimate Goal

Give students the right tools and they will excel! When students gain confidence in math, their higher-level thinking blossoms. Understanding multiplication and division is essential for strong math skills and Sticks and Steps® will surely be an enjoyable part of their math journey.

LaVergne, TN USA
26 January 2011
213976LV00001B/2/P